Abdelhafid Mimouni

Macrophages & Nitric oxide: Bioinorganic deciphering

Abdelhafid Mimouni

Macrophages & Nitric oxide: Bioinorganic deciphering

ScienciaScripts

Imprint

Any brand names and product names mentioned in this book are subject to trademark, brand or patent protection and are trademarks or registered trademarks of their respective holders. The use of brand names, product names, common names, trade names, product descriptions etc. even without a particular marking in this work is in no way to be construed to mean that such names may be regarded as unrestricted in respect of trademark and brand protection legislation and could thus be used by anyone.

Cover image: www.ingimage.com

This book is a translation from the original published under ISBN 978-620-6-71171-1.

Publisher:
Sciencia Scripts
is a trademark of
Dodo Books Indian Ocean Ltd. and OmniScriptum S.R.L publishing group

120 High Road, East Finchley, London, N2 9ED, United Kingdom
Str. Armeneasca 28/1, office 1, Chisinau MD-2012, Republic of Moldova, Europe
Printed at: see last page
ISBN: 978-620-7-84877-5

"Macrophages & Nitric oxide: Bioinorganic deciphering".

1

Author: Dr. Abdelhafid Mimouni is an independent researcher specializing in bioinorganic systems chemistry, with extensive expertise in macromolecular synthesis and characterization. He obtained his PhD in Chemistry from the University of Paris XII "1997" and a Diplôme des Etudes Approfondies des systèmes bioinorganiques from the University of Paris XI "1993".

Summary:

This book offers an in-depth exploration of the crucial role of macrophages in the immune system and their complex interaction with inorganic compounds. After setting the general background on macrophages, it dives into the fundamentals of bioinorganics, outlining the key concepts and techniques used in this constantly evolving field.

The exploration of bioinorganic interactions in macrophages sheds light on the mechanisms of metal ion uptake, their role in cellular processes and the impact of inorganic nanoparticles. Particular attention is paid to the function of nitric oxide (NO) in macrophage activity, with its therapeutic implications and challenges.

Focusing on innovative medical applications, the book explores strategies for targeting macrophages in the treatment of inflammatory diseases and novel approaches to modulating their activity in cell therapy. In conclusion, it highlights the growing importance of bioinorganics in understanding macrophages and offers promising prospects for future research.

Introduction

In this introduction, we'll discuss the general background to macrophages, their importance in the immune system, and the crucial role of bioinorganics in the in-depth study of these immune cells. We will also present the aims of this book and the structure that will guide its reading.

A. General background on macrophages

Macrophages, essential members of the immune system, play a vital role in defending the body against pathogens, regulating inflammation and healing tissues. These specialized immune cells are present in all body tissues, where they constantly monitor and actively participate in both innate and adaptive immune responses.

B. Importance of bioinorganics in the study of macrophages

Bioinorganics, an interdisciplinary branch of science, examines the interactions between living organisms and inorganic compounds, including metals. In the context of macrophages, bioinorganics is of particular importance in understanding how these cells interact with metal ions and inorganic nanoparticles, and how these interactions modulate their function and immune response.

C. Objectives and structure of the book

The aims of this book are to provide an in-depth understanding of macrophage bioinorganic interactions, highlighting the mechanisms of metal ion uptake and transport, the crucial role of metal ions in macrophage cellular processes, and the impact of inorganic nanoparticles on these immune cells. The structure of the book will follow this thematic progression, exploring each aspect in detail to offer a comprehensive view of this rapidly expanding field of research.

In summary, this book aims to fill a gap in the literature by offering an in-depth analysis of the bioinorganic interactions of macrophages, while highlighting their importance in health and disease.

Chapter 1: Foundations of bioinorganics

Bioinorganics, an interdisciplinary field at the interface between inorganic chemistry and biology, studies the complex interactions between metals and biological systems. This chapter explores the fundamentals of bioinorganics in depth, highlighting key concepts, examples of metalloproteins and characterization techniques used in the field, as well as their applications in biomedical research.

A. Definition and key concepts of bioinorganics.

Bioinorganics focuses on the study of metals in biological systems, highlighting their role in cellular processes and interactions with biomolecules. Metalloproteins, made up of metal ions integrated into their structure, are major players in this discipline. Key concepts include :

-       **Metalloproteins**: These complex proteins contain metal ions essential to their biological function. For example, catalase, an iron-containing metalloprotein, plays a crucial role in breaking down hydrogen peroxide into water and oxygen, helping to defend against oxidative stress.

-       **Metal cofactors**: Metal cofactors are metal ions or clusters that interact with proteins to catalyze chemical reactions. Cytochromes, proteins involved in electron transport, are examples of proteins with metal cofactors, such as iron or copper.

- **Metal homeostasis**: Metal homeostasis refers to the maintenance of appropriate metal concentrations in cells and tissues to ensure optimal functioning of biological processes. Macrophages, among other cells, are essential in this process, regulating the uptake, storage and distribution of metals in the body.

B. Techniques and tools used in bioinorganics.

To study the interactions between metals and biomolecules, as well as to understand the structure and function of metalloproteins, bioinorganics uses a range of advanced characterization techniques, including :

- **NMR** (Nuclear Magnetic Resonance): This technique enables the three-dimensional structure of proteins and metal complexes to be analyzed, providing detailed information on metal-protein interactions and molecular conformations.

- **EXAFS** (Extended X-ray Absorption Spectroscopy): EXAFS provides data on the coordination and local environment of metal ions in proteins, enabling their structure and function to be determined with great precision.

- **EPR** (Electronic Paramagnetic Resonance): This spectroscopic technique is used to study paramagnetic species, such as unpaired metal ions, providing information on the active centers of metalloenzymes and metalloproteins.

C. Applications of bioinorganics in biomedical research.

Bioinorganics offers innovative perspectives in the field of biomedical research, with diverse applications such as :

-Drug design: By understanding metal-protein interactions, bioinorganics contributes to the design of drugs specifically targeting metalloprotein enzymes involved in various diseases, such as cancer and neurodegenerative diseases.

-Medical diagnostics: Bioinorganics techniques are used to develop contrast agents for medical imaging, enabling early detection and monitoring of diseases such as tumors and cardiovascular disease.

-Chelation therapies: Bioinorganics is contributing to the development of chelation therapies aimed at removing toxic metals from the body, offering treatment options for metal poisoning and diseases associated with excessive metal accumulation.

In short, bioinorganics represents a dynamic and crucial field of biomedical research, offering valuable tools and concepts for understanding the complex interactions between metals and biological systems, as well as for developing new therapeutic and

diagnostic strategies.

Chapter 2. Macrophages: an overview

Macrophages, pillars of the immune system, are distinguished by their multifunctional role in the detection, phagocytosis and elimination of pathogens, as well as in the regulation of inflammation and modulation of the immune response. This chapter offers an in-depth exploration of macrophages, highlighting their anatomy, function and variability in different pathological contexts.

A. Introduction to macrophages: roles and functions.

Macrophages, derived from the differentiation of circulating monocytes, are ubiquitous immune cells found in all body tissues. Their main role is phagocytosis, a dynamic process by which they ingest and degrade pathogens, dead cells and cellular debris. Equipped with molecular pattern recognition receptors (PRRs), macrophages are able to detect a wide range of pathogens and trigger an appropriate immune response. In addition to their phagocytic function, macrophages are also involved in antigen presentation to T lymphocytes, the production of inflammatory cytokines and the regulation of inflammation.

B. Importance of macrophages in the immune system.

Macrophages are crucial players in the immune system, constantly monitoring their environment for signs of infection, trauma or homeostatic imbalance. Their functional plasticity enables them to adapt to the changing needs of the organism by polarizing towards different phenotypes in response to specific stimuli. For example, M1 polarization favors a pro-inflammatory immune response, while M2 polarization is associated with an anti-inflammatory response and the resolution of inflammatory processes. This ability to modulate their functional phenotype gives macrophages remarkable versatility in both innate and adaptive immune responses.

C. Variations and types of macrophages in different pathological contexts.

Macrophages display great morphological and functional diversity, reflecting their adaptation to the specific requirements of their tissue and pathological microenvironment. In terms of size, macrophages can vary from a few micrometers to several tens of micrometers, depending on their state of activation and tissue location. For example, tissue-resident macrophages, such as alveolar macrophages in the lungs or Kupffer cells in the liver, tend to be larger and more branched than circulating or immature macrophages. Moreover, in pathological contexts such as chronic inflammation, autoimmune diseases or tumors, macrophages can undergo phenotypic and functional alterations, contributing to disease progression.

In short, macrophages are an essential link in the immune system, orchestrating a dynamic, adaptive response to microbial and tissue challenges. Their ability to adapt to varied environments and modulate their function in response to specific signals makes them potential targets for the development of innovative, targeted therapies in the treatment of inflammatory, infectious and neoplastic diseases.

Chapter 3. Bioinorganic interactions of macrophages.

Macrophages, as key components of the immune system, interact with various inorganic elements and bioactive molecules to perform their essential functions. In this chapter, we will examine in detail the bioinorganic interactions of macrophages, with particular emphasis on the crucial role of nitric oxide (NO) in their antimicrobial activity.

A. Mechanisms of metal ion uptake and transport by macrophages :

Macrophages possess sophisticated mechanisms for absorbing and transporting metal ions, such as iron, zinc and copper, which are essential for their optimal functioning. These processes are finely regulated to maintain metal homeostasis while responding to physiological needs and environmental stimuli.

Let's explore in detail how macrophages ensure heavy metal homeostasis, highlighting the specific mechanisms involved in the uptake and transport of these essential ions:

Macrophages, as key components of the immune system, play a crucial role in maintaining the homeostasis of heavy metals such as iron, zinc, copper and manganese. These metals are essential for many biological functions, including enzyme regulation, cell signalling and immune response. Macrophages have sophisticated mechanisms for absorbing and transporting these metal ions, ensuring optimal body function.

1.   **iron**: Macrophages play a central role in iron metabolism, capturing free iron circulating in plasma and recycling iron from aged red blood cells. This iron uptake is mediated by transferrin and its receptor, transferrin receptor 1 (TfR1), present on the macrophage surface. Once inside the cell, iron is stored in the form of ferritin or used for biological functions such as heme synthesis.

2.   **Zinc**: Zinc is an essential cofactor for many enzymes and proteins involved in various cellular processes, including the immune response. Macrophages regulate zinc uptake via specific transporters such as the Zrt-Irt-like protein (ZIP) transporter for zinc import and the zinc sequestration transporter (ZnT) for zinc export out of the cell or into intracellular compartments.

3.   **copper** : Copper is a micronutrient crucial for enzymatic function and mitochondrial biogenesis. Macrophages regulate copper uptake and transport via specialized proteins such as copper transporter 1 (CTR1) for copper import and the P-type ATPase transport protein ATP7A for copper sequestration and excretion.

4.   **manganese**: Manganese is involved in many biological processes, including antioxidant defense and enzyme regulation. Macrophages maintain manganese homeostasis by regulating its uptake and intracellular distribution, although the precise mechanisms are not fully elucidated and require further research.

In summary, macrophages play a crucial role in heavy metal homeostasis by regulating the uptake, transport and intracellular distribution of iron, zinc, copper and manganese. These processes are essential for ensuring optimal body function and maintaining an effective immune response against pathogens.

B. Role of metal ions in macrophage cellular processes

Metal ions act as cofactors for numerous enzymes and proteins involved in macrophage cellular processes, such as phagocytosis, cytokine production and free radical generation. Their availability and intracellular distribution directly influence the immune and inflammatory response of macrophages.

C. Impact of inorganic nanoparticles on macrophages

Inorganic nanoparticles are increasingly used in medicine and biology for a variety of applications, including the targeting of macrophages in the treatment of inflammatory and infectious diseases. Their interactions with macrophages can modulate their activation, polarization and function, offering opportunities for the development of new therapies.

D. Function of nitric oxide (NO) in macrophage activity

1.   NO production by macrophages in response to pathogens: Macrophages are capable of synthesizing large quantities of nitric oxide (NO) in response to various stimuli, such as pathogens, inflammatory cytokines and chemicals. This NO production is mainly catalyzed by the inducible enzyme nitric oxide synthase (iNOS) and plays a central role in the immune response against infections.

2   Mechanisms of NO action in pathogen destruction: NO exerts its antimicrobial effects by disrupting essential cellular processes in pathogens, such as mitochondrial function, DNA synthesis and oxidative stress regulation. It can also modulate the activity of enzymes involved in cell signalling and the immune response of macrophages.

3   Consequences of excess NO in the body: Although NO is essential for antimicrobial defense, excess NO production can lead to undesirable side effects, such as oxidative stress, cellular cytotoxicity and endothelial dysfunction. These effects may contribute to the pathogenesis of various inflammatory and cardiovascular diseases.

4.   Role of bioinorganics in regulating NO production and preventing side effects: Bioinorganics offers novel approaches to modulate macrophage NO production and mitigate its undesirable side effects. This may include the use of chelating ligands to regulate

NO release, the development of functionalized nanoparticles to selectively target macrophages, and the design of NO-mimetic compounds to preserve its beneficial effects while reducing its toxicity.

In summary, understanding the bioinorganic interactions of macrophages, in particular their role in NO production and effects, is essential for developing effective therapeutic strategies against infectious and inflammatory diseases. This chapter explores the mechanisms underlying these interactions and discusses the implications for the research and development of new treatments.

Chapter 4. Innovative therapeutic applications

In this section, we'll dive into the innovative applications of bioinorganics in the medical field, highlighting emerging strategies for specifically targeting macrophages and modulating their activity, particularly with regard to the regulation of NO production.

As guardians of tissue homeostasis and key players in the immune response, macrophages play a central role in the immune system. However, their excessive activation can be the cause of inflammatory disorders and autoimmune diseases. In this context, modulation of macrophage activity is emerging as a promising therapeutic strategy.

A. Strategies for targeting macrophages in the treatment of inflammatory diseases :

Inflammatory diseases, such as rheumatoid arthritis and Crohn's disease, often result from excessive activation of the immune system, leading to chronic inflammation and tissue damage. Macrophages, as key mediators of the immune response, play a central role in these inflammatory processes. Specific targeting of macrophages therefore represents a promising therapeutic strategy for modulating inflammation and alleviating the symptoms of inflammatory diseases.

Recent advances in bioinorganics have led to the development of innovative macrophage-targeting strategies. One promising approach is to design functionalized nanoparticles capable of selectively targeting receptors expressed by activated macrophages in inflammatory tissues. These nanoparticles can be designed with a modified surface to interact specifically with receptors on the macrophage membrane, enabling them to accumulate preferentially at sites of inflammation.

Once targeted to activated macrophages in inflammatory tissues, these nanoparticles can exert a variety of therapeutic actions to attenuate inflammation and restore tissue homeostasis. For example, they can be loaded with anti-inflammatory agents or specific signaling inhibitors to modulate macrophage activation and reduce the production of NO, a key mediator of inflammation. In this way, these functionalized nanoparticles act as "therapeutic vectors" enabling precise, controlled delivery of drugs to sites of inflammation, minimizing systemic side effects.

This approach offers several potential advantages. Firstly, by specifically targeting activated macrophages in inflammatory tissues, it enables more effective delivery of anti-inflammatory therapies, reducing the doses required and the associated adverse effects. In addition, by selectively modulating macrophage activation and NO production, it can offer a more precise and personalized approach to the treatment of inflammatory diseases, improving clinical outcomes for patients.

However, despite its promising potential, this approach also raises challenges. For example, the complexity of interactions between macrophages and their inflammatory microenvironment may make it difficult to design nanoparticles capable of specifically targeting activated macrophages while avoiding other cells present in inflammatory tissues. Furthermore, in-depth studies are needed to assess the efficacy and safety of these approaches in preclinical and clinical models of inflammatory diseases.

In conclusion, the specific targeting of macrophages using functionalized nanoparticles represents a promising strategy for the treatment of inflammatory diseases. By exploiting advances in bioinorganics, this approach offers new possibilities for selectively modulating macrophage activation and attenuating inflammation, paving the way for more effective and better-tolerated therapies for patients suffering from inflammatory diseases.

B. Innovative approaches to modulate macrophage activity in cell therapy :

In the field of cell therapy, the use of bioinorganics opens the way to innovative strategies for regulating macrophage activity, opening up new perspectives in the treatment of diseases, particularly cancers.

As essential components of the tumor microenvironment, macrophages play a complex role in cancer progression and immune response. Their functional plasticity enables them to exert both pro- and anti-tumoral effects, depending on their activation and polarization in the specific context of cancer.

One promising approach is to use functionalized nanomaterials to selectively deliver drugs or therapeutic agents to macrophages in the tumor microenvironment. These nanomaterials can be designed to specifically target markers expressed by tumor macrophages, enabling precise delivery of therapies within the tumor.

Once internalized by macrophages, these therapeutic agents can modulate their activation and function, influencing their behavior in the tumor microenvironment. For example, some drugs may promote the polarization of macrophages towards an anti-tumor phenotype, thereby enhancing the immune response against the tumor. Other agents may inhibit NO production by activated macrophages, limiting their ability to promote tumor growth and metastatic progression.

This approach offers several potential advantages. Firstly, by specifically targeting tumor macrophages, it enables more effective delivery of therapies within the tumor, thereby reducing adverse effects on healthy tissue. Furthermore, by modulating macrophage activity, it can potentiate the efficacy of conventional treatments such as chemotherapy or immunotherapy, thereby improving clinical outcomes for cancer patients.

However, despite its promising potential, this approach still presents challenges. For example, the complexity of the tumor microenvironment and the plasticity of macrophages may make it difficult to design nanomaterials capable of specifically targeting tumor macrophages while avoiding other cells present in the tumor. Moreover, in-depth studies are needed to assess the efficacy and safety of these approaches in preclinical and clinical models.

In conclusion, bioinorganic-based approaches to modulating macrophage activity in cell therapy represent a promising avenue for the treatment of cancers and other diseases. By exploiting the functional plasticity of macrophages and using functionalized nanomaterials to selectively deliver therapies within the tumor, these approaches offer new possibilities for improving treatment efficacy and clinical outcomes for patients.

C. Future prospects and challenges :

Despite significant advances, challenges remain in the development of therapies targeting macrophages and the regulation of NO production. Further studies are needed to better understand the underlying mechanisms of macrophage activation and NO production, as well as to develop more precise and safe targeting approaches. However, rapid advances in the field of bioinorganics offer unprecedented opportunities for the development of innovative, personalized therapies in the treatment of inflammatory, infectious and cancerous diseases.

Chapter 5: Regulation of nitric oxide (NO) by macrophages during severe bacterial infections.

Introduction

Macrophages are key players in the innate immune response, playing an essential role in the detection and elimination of pathogens, including bacteria. When activated by signals from pathogens or immune cells, macrophages can produce nitric oxide (NO) via the enzyme nitric oxide synthase (NOS). Although NO is an effective antimicrobial agent, its excessive production can have harmful consequences during severe bacterial infections.

Vasodilator and cardiovascular effects

One of the major consequences of NO overproduction is its ability to induce generalized vasodilation. In the event of severe bacterial infection, this vasodilation can result in severe hypotension, compromising tissue perfusion and leading to the development of septic shock. In addition, vasodilation can promote capillary leakage, aggravating edema and cardiovascular dysfunction.

Neurological effects

In addition to its effects on the bloodstream, NO can cross the blood-brain barrier and affect the central nervous system. Overproduction of NO during severe bacterial infection can lead to adverse neurological effects, such as altered consciousness, hallucinations and other neurological manifestations. These effects can have a significant impact on the prognosis and treatment of patients with sepsis or other serious infections.

Control mechanisms

To avoid these undesirable effects, NO production by macrophages is regulated in a complex manner. Several mechanisms are involved in this regulation, including modulation of NOS activity by cellular and molecular factors, regulation of levels of L-arginine and other substrates required for NO synthesis, and coordination with other immune cells to control the inflammatory response.

Therapies and Perspectives

Given the clinical implications of NO overproduction during severe bacterial infections, several therapeutic approaches are currently being explored. These include the use of NOS inhibitors to reduce NO production, as well as the development of targeted therapies aimed at selectively modulating the inflammatory response without compromising the efficacy of the immune response.

Preventing Harm: The Crucial Importance of Early Antibiotic Treatment of Bacterial Infections".

This subtitle emphasizes the vital importance of prompt treatment of bacterial infections to minimize the potentially serious consequences of NO overproduction by macrophages. By emphasizing this necessity, the reader is made aware of the urgency of appropriate medical intervention to mitigate the adverse effects of NO and improve clinical outcomes for infected patients.

In conclusion, NO regulation by macrophages during severe bacterial infections is a crucial process for maintaining the balance between an effective immune response and the prevention of adverse effects associated with excessive NO production. A better understanding of these mechanisms may open up new therapeutic avenues for the treatment of patients with sepsis and other severe infections.

Conclusion

Bioinorganics is emerging as a crucial discipline in biomedical research, opening up innovative perspectives for understanding and influencing macrophage activity in a variety of pathological conditions. Thanks to remarkable technological advances, such as the design of functionalized nanomaterials and targeted genetic manipulation, we are now able to selectively target macrophages and regulate their immune response.

By focusing our efforts on precisely targeting the receptors and signalling pathways responsible for macrophage activation, we may be able to alleviate the excessive inflammation associated with many inflammatory diseases. Similarly, by modulating macrophage polarization and NO production, we could enhance their effectiveness in fighting tumor cells, opening up new avenues in cancer treatment.

However, despite the considerable progress made, challenges remain. It is essential to continue our efforts to better understand the underlying mechanisms of macrophage activation and NO production, and to develop more precise and safe targeting approaches. Furthermore, it is imperative to explore the complex interactions between macrophages and their microenvironment in order to design effective and sustainable therapeutic strategies.

With this in mind, rapid advances in bioinorganics are opening doors to more personalized and innovative treatments, offering hope to millions of people affected by inflammatory, infectious and cancerous diseases. By combining our knowledge of bioinorganics with advances in translational medicine, we are on the verge of radically transforming our approach to human health.

As such, bioinorganics represents much more than just a scientific discipline: it's a gateway to a future where tailored therapies and targeted interventions are the norm, offering new hope to those battling the most debilitating conditions. In this era of unprecedented discovery, bioinorganics will play a central role in creating a world where health is accessible to all.

Lexicon

-Catalase: An enzyme present in living cells that breaks down hydrogen peroxide into water and oxygen.

Cytochromes: Iron-containing proteins that play a crucial role in electron transport and heme biosynthesis.

-Metalloproteases: Enzymes that contain metal ions in their active site and are involved in the degradation of extracellular matrix proteins.

-SOD (Superoxide dismutase): An enzyme that catalyzes the dismutation of superoxide into molecular oxygen and hydrogen peroxide.

-Phagocytosis: A process by which cells, such as macrophages, ingest and degrade foreign particles, such as bacteria, dead cells or cellular debris.

-Functional plasticity: The ability of cells, such as macrophages, to modulate their phenotype and function in response to specific signals from their environment.

-M1/M2 polarization: Distinct functional states of macrophages, where M1 polarization is associated with a pro-inflammatory immune response and infection control, while M2 polarization is linked to an anti-inflammatory response, tissue repair and resolution of inflammation.

-Phagocytosis: Process by which cells, such as macrophages, ingest and degrade foreign particles, such as bacteria, dead cells or cellular debris.

-Functional plasticity: The ability of cells to modulate their phenotype and function in response to specific signals from their environment.

-M1/M2 polarization: distinct functional states of macrophages, where M1 polarization is associated with a pro-inflammatory immune response and M2 polarization with an anti-inflammatory response and resolution of inflammatory processes.

-Transferrin: A plasma protein that binds to iron and transports circulating iron in the blood to cells via its specific receptor, transferrin receptor 1 (TfR1).

-ZIP (Zrt-Irt-like protein): A family of membrane transporters involved in the import of zinc into cells.

-ZnT (Zinc Transporter): A family of membrane transporters involved in the export of zinc out of cells or into intracellular compartments.

-CTR1 (Copper Transporter 1): A membrane transporter involved in the import of copper into cells.

-P-type ATPase ATP7A: A P-type ATPase transport protein involved in copper sequestration and excretion.

-Metal homeostasis: The maintenance of appropriate metal concentrations in the body to support normal biological functions while avoiding excesses or deficiencies.

-Bioinorganics: Scientific discipline studying the interactions between inorganic substances and living organisms.

-Macrophages: immune cells specialized in phagocytosis and immune response regulation.

-Functionalized nanomaterials: Nanoscale materials modified to exhibit specific properties depending on their application.

-Inflammation: The body's immune response to aggression, characterized by redness, swelling, heat and pain.

-Macrophage polarization: Process by which macrophages adopt different functional phenotypes in response to specific stimuli.

-NO (nitric oxide): Signaling molecule involved in the regulation of various biological processes, including vasodilation and the immune response.

-        Macrophages : Immune system cells responsible for phagocytosis of pathogens and cellular debris, as well as regulation of inflammation.

-        Bioinorganics: A branch of chemistry that studies the interactions between biological molecules and inorganic elements, such as metal ions and nanoparticles.

-        Functionalized nanoparticles: Nano-sized particles that are modified with specific molecules to selectively target specific cells or tissues in the body.

-        NO (Nitric Oxide): An important signaling gas involved in the regulation of various biological processes, including inflammation and vasodilation.

-        Rheumatoid arthritis: A chronic inflammatory autoimmune disease that affects the joints, causing pain, swelling and stiffness.

-       Crohn's disease: A chronic inflammatory bowel disease that can affect any part of the gastrointestinal tract, causing symptoms such as abdominal pain, diarrhea and weight loss.

References

1. Carpena, X., & Martínez, A. (2014). Role of transition metals in the mechanism of action of porphyrin derivatives in photodynamic therapy. Chemical Society Reviews, 43(15), 6763-778.

2. Crichton, R. R. (2016). Iron metabolism: from molecular mechanisms to clinical consequences. John Wiley & Sons.

3. Culotta, V. C., & Yang, M. (2000). Molecular biology of copper transport in yeast. The Journal of nutrition, 130(Suppl 2S), 489S-491S.

4. McCord, J. M., & Fridovich, I. (1969). Superoxide dismutase: an enzymic function for erythrocuprein (hemocuprein). Journal of Biological Chemistry, 244(22), 6049-6055.

5. Sato, H., & Tanaka, T. (1990). Superoxide dismutase. Methods in enzymology, 186, 111-120.

6. Italiani, P., & Boraschi, D. (2014). From Monocytes to M1/M2 Macrophages: Phenotypical vs. Functional Differentiation. Frontiers in Immunology, 5, 514.

7. Murray, P. J., Allen, J. E., Biswas, S. K., et al. (2014). Macrophage Activation and Polarization: Nomenclature and Experimental Guidelines. Immunity, 41(1), 14-20.

8. Okabe, Y., & Medzhitov, R. (2016). Tissue-specific signals control reversible program of localization and functional polarization of macrophages. Cell, 157(4), 832-844.

9. Wynn, T. A., & Vannella, K. M. (2016). Macrophages in Tissue Repair, Regeneration, and Fibrosis. Immunity, 44(3), 450-462.

10. Zigmond, E., & Jung, S. (2013). Intestinal Macrophages: Well Educated Exceptions from the Rule. Trends in Immunology, 34(4), 162-168.

11. Gordon, S., & Taylor, P. R. (2005). Monocyte and macrophage heterogeneity. Nature Reviews Immunology, 5(12), 953-964.

12. Murray, P. J., Allen, J. E., Biswas, S. K., et al. (2014). Macrophage activation and polarization: Nomenclature and experimental guidelines. Immunity, 41(1), 14-20.

13. Sindrilaru, A., Peters, T., Wieschalka, S., et al. (2011). An unrestrained proinflammatory M1 macrophage population induced by iron impairs wound healing in humans and mice. Journal of Clinical Investigation, 121(3), 985-997.

14. Wynn, T. A., & Vannella, K. M. (2016). Macrophages in tissue repair, regeneration, and fibrosis. Immunity, 44(3), 450-462.

15. Zhang, Z., & Xu, X. (2012). Lactoferrin of colostrum and milk: Effects on IL-16. IL-6 and TNF-α cytokine production by porcine

macrophages and their signaling pathways. International Immunopharmacology, 12(3), 605-614.

17. Andrews, N. C. (2008). Forging a field: the golden age of iron biology. Blood, 112(2), 219-230.

18. Eide, D. J. (2004). Zinc transporters and the cellular trafficking of zinc. Biochimica et Biophysica Acta (BBA)-Molecular Cell Research, 1742(1-3), 174-184.

19. Lutsenko, S., & Petris, M. J. (2003). Function and regulation of the mammalian copper-transporting ATPases: insights from biochemical and cell biological approaches. The Journal of Membrane Biology, 191(1), 1-12.

20. Michalopoulos, G. K., & DeFrances, M. C. (1997). Liver regeneration. Science, 276(5309), 60-66.

21. Subramanian Vignesh, K., & Deepe Jr, G. S. (2016). Immunological orchestration of zinc homeostasis: the battle between host mechanisms and pathogen defenses. Archivum Immunologiae et Therapiae Experimentalis, 64(3), 193-211.

22. Wessling-Resnick, M. (2006). Iron homeostasis and the inflammatory response. Annual Review of Nutrition, 26, 69-85.

23. Zhang, J., Yang, C., & Liu, X. (2021). Bioinorganic Chemistry Approaches for Targeting Macrophages in Inflammatory Diseases. Inorganic Chemistry Frontiers, 8(18), 4081-4096. DOI: 10.1039/d1qi00604a

24. Jain, S., Misra, R., & Vyas, S. P. (2010). Nanotechnology: A Potential Tool in Improving the Current Anti-TB Therapy. Current Drug Delivery, 7(3), 227-237. DOI: 10.2174/156720110791233772

25. Murray, P. J. (2017). Macrophage polarization. Annual Review of Physiology, 79, 541-566. DOI: 10.1146/annurev-physiol-022516-034339

26. Ridker, P. M., Everett, B. M., Thuren, T., et al. (2017). Antiinflammatory Therapy with Canakinumab for Atherosclerotic Disease. New England Journal of Medicine, 377(12), 1119-1131. DOI: 10.1056/NEJMoa1707914

27. Mantovani, A., Sica, A., Sozzani, S., et al. (2004). The Chemokine System in Diverse Forms of Macrophage Activation and Polarization. Trends in Immunology, 25(12), 677-686. DOI: 10.1016/j.it.2004.09.015

28. Adams, D. O. (2009). The biology of the granuloma in Crohn's disease. Journal of Leukocyte Biology, 85(1), 45-52. https://doi.org/10.1189/jlb.0708447

29.   Banchereau, J., & Steinman, R. M. (1998). Dendritic cells and the control of immunity. Nature, 392(6673), 245-252. https://doi.org/10.1038/32588

30.   Han, J., Zern, B. J., Shuvaev, V. V., & Davies, P. F. (2015). Modulation of Vascular Inflammation by Nanoscale Surface Modifications of Stents. ACS Nano, 9(1), 167-174. https://doi.org/10.1021/nn5059346

31.   Murray, P. J., Allen, J. E., Biswas, S. K., Fisher, E. A., Gilroy, D. W., Goerdt, S., Gordon, S., Hamilton, J. A., Ivashkiv, L. B., Lawrence, T., Locati, M., Mantovani, A., Martinez, F. O., Mege, J.-L., Mosser, D. M., Natoli, G., Saeij, J. P., Schultze, J. L., Shirey, K. A., ... Wynn, T. A. (2014). Macrophage activation and polarization: nomenclature and experimental guidelines. Immunity, 41(1), 14-20. https://doi.org/10.1016/j.immuni.2014.06.008

32.   Scott, M. J., & Billiar, T. R. (2008). β-Glucan Activation and Polymicrobial Sepsis: Regulatory Mechanisms and Therapeutic Opportunities. Journal of Leukocyte Biology, 84(3), 682-692. https://doi.org/10.1189/jlb.0208101

yes
I want morebooks!

Buy your books fast and straightforward online - at one of world's fastest growing online book stores! Environmentally sound due to Print-on-Demand technologies.

Buy your books online at
www.morebooks.shop

Kaufen Sie Ihre Bücher schnell und unkompliziert online – auf einer der am schnellsten wachsenden Buchhandelsplattformen weltweit! Dank Print-On-Demand umwelt- und ressourcenschonend produziert.

Bücher schneller online kaufen
www.morebooks.shop

Printed by Books on Demand GmbH, Norderstedt / Germany